Y0-EGI-676

EXPLORING WITH SOLAR ENERGY

EXPLORING
WITH

Illustrated with photographs and drawings

SOLAR ENERGY

Thomas H. Metos

and Gary G. Bitter

Julian Messner New York

Published by Julian Messner, a Simon & Schuster
Division of Gulf & Western Corporation,
Simon & Schuster Building,
1230 Avenue of the Americas,
New York, N.Y. 10020.

JULIAN MESSNER and colophon are trademarks of Simon & Schuster, registered in the U.S. Patent and Trademark Office.

Manufactured in the United States of America

Design by Sheila Lynch

Third Printing, 1980

Picture credits

Arizona Public Service, p. 28

Gary G. Bitter, pp. 35, 37 bottom, 39, 41, 46, 47, 49

Copper Development Association, Inc., p. 25

William Jaber, pp. 32, 40, 44 top, 50

Jeffery T. Metos, pp. 9, 13, 23, 24, 26, 29, 31, 33, 37 top, 41 right, 42, 43, 44 bottom, 45, 55

Bruno J. Rolak, Command Historian, Fort Huachuca, Arizona, p. 14

Smithsonian Institution, pp. 16, 17, 18, 19, 41 left

Wide-World Photos, Inc., p. 61

Library of Congress Cataloging in Publication Data

Metos, Thomas H
Exploring with solar energy.

Includes index.
SUMMARY: Discusses the historical, present, and future uses of solar energy and includes instructions for experiments and model building.
1. Solar energy—Juvenile literature. [1. Solar energy] I. Bitter, Gary G., joint author. II. Title.
TJ810.M47 333.7 78-15179
ISBN 0-671-32948-0

CONTENTS

What is Solar Energy? 7

The Sun 8

Early Use of Solar Energy 11

Using Solar Energy Today 20

Experimenting with Solar Energy 30

The Future and Solar Energy 56

Index 63

To Marilyn, Melissa, and Jeffery
and
To Kay, Steven, Michael, and Matthew
–and a special thanks to Mady

WHAT IS SOLAR ENERGY?

"OUCH, IT'S HOT," said Jerry, opening the car door. The handle burned, and a rush of hot air hit her face. The family car had been sitting in the shopping center parking lot for several hours this bright summer day, and Jerry had just experienced a form of solar energy—the heat of the sun.

You've probably had something of the same kind happen to you. Or perhaps you have experienced another form of solar energy—certain light rays from the sun that can burn your skin.

Solar energy is the light and heat that comes from the sun. This light and heat is made up of electromagnetic waves such as X rays, infrared rays, ultraviolet rays and gamma rays. Most of the sun's energy that reaches the earth is in the form of light—infrared and ultraviolet light. All of the sun's energy reaching the earth is called *radiation*.

THE SUN

The sun is a star at the center of our solar system. It is the brightest of all stars to our eyes, because it is closest to earth, some 93,000,000 miles away.

Astronomers call the sun an orange dwarf star, although it is not a dwarf-size star.

The sun is enormously hot. The temperature at the core, or center of the sun is over 27,000,000° F. The sun's surface temperature is 10,000° F.

The sun is the source of almost all of our life and energy on earth. Without it, the earth would be a dark and barren rock floating through space.

Without the sun we would have no food, no trees, no weather. We would have no natural or direct light or heat,

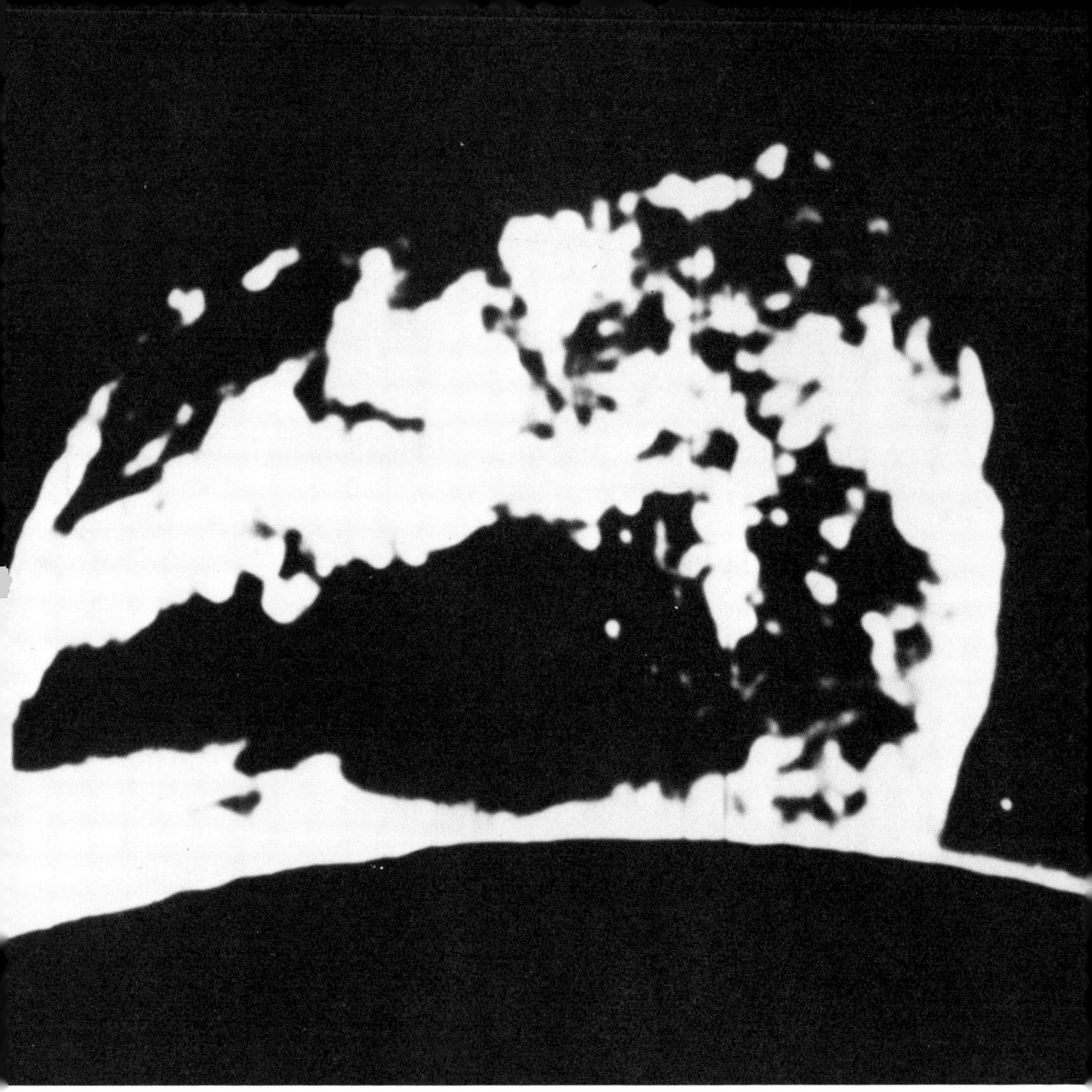

A solar flare—a great outburst of energy from the sun's surface.

no fossil fuels to make artificial light and heat. The *fossil fuels*—oil, gas, and coal—were once either plants or animals that were nourished by the sun, died, and through millions of years were transformed by pressure and heat into these fossil fuels. Solar energy has created these sources of energy for our planet. The sun is still creating fossil fuels, but we are using them up faster than they are being made.

Because of the increasing shortage of fossil fuels, the world is looking to other sources of energy, like nuclear energy, and solar energy, to supply our homes, industries, and means of transportation.

The earth receives only one one-thousandth of a millionth of the sun's energy. There are several reasons for this: the distance we are from the sun, the clouds, and the reflection of the light by the earth's *atmosphere*, or air. Still, the earth receives enough energy from the sun *every three days* to equal *all* of our remaining stored fossil fuels.

Use of the sun's power to furnish energy has been known for thousands of years. Fossil fuels and hydroelectric power were cheap and plentiful, however, so most uses of solar energy were discarded. But now we are finally trying to harness the power of the sun.

EARLY USE OF SOLAR ENERGY

Human beings from their beginning on earth have needed the sun's heat and light to survive. Many groups, aware of their dependency on the sun, became sun worshipers. The Egyptians had a sun god, Ra. The Greeks called theirs Helios. The Aztec sun god was named Tezcatlipoca (tez-kot-lee-poh-kah), which means "smoking mirror." There were many others as well.

An *eclipse* of the sun, when the shadow of another heavenly body passes in front of it, terrified people, who felt the gods were angry with them, or that the sun god was fighting with other gods.

One of the earliest uses of solar energy is described in the legend of Archimedes and the burning mirrors at the Battle of Syracuse in 212 B.C. Archimedes, a famous scientist, is said to have made the soldiers use their highly polished shields to reflect the sun's rays on the attacking Roman fleet to set the wooden ships afire. Syracuse was saved from the invading Romans. Today, even though we know that small "burning mirrors" were well known at the time, some scientists believe that Archimedes did not use burning mirrors. It would have been most difficult to keep the sun's rays focused on one place on each ship long enough to set it afire.

The ancient Indians produced one of the finest examples of naturally heated and cooled buildings ever constructed in this country. Located in Arizona, it is called Montezuma's Castle. The early pioneers mistakenly believed it had been built by Aztec Indians fleeing from Mexico after the Spanish conquest, and named it after Montezuma, emperor of the Aztecs. The building is in the center of a white limestone cliff, with a large overhang of rock on the top. The seven-story structure of *adobe*, sun-dried bricks of clay and straw, receives the full force of the sun in the winter for warmth. The building and the surrounding rocks hold the heat after the sun has set, so the building remains warm at night. During the summer, the building is completely shaded by the rocks, and the wind further cools the structure.

The sun's power was also used by the United States Army in the Indian wars in the west. A chain of heliograph stations, set up on New Mexico and Arizona mountain

The ancient Native American ruins called Montezuma's Castle.

Signalmen engaged in heliograph training at Camp Wikoff, New York, in 1898.

A MANCE heliograph, the type used during the Geronimo campaign of 1886.

tops, helped General Nelson A. Miles to keep track of the warring Apache Indians. This led to their defeat in 1886. *Heliographs* were mirrors mounted on tripods. A shutter on the heliograph allowed the operator to give short or long flashes of reflected sun light and thus use the Morse Code to send light messages over 50 or 60 miles.

Solar power has also been used over the ages to dry and cure crops. A number of crops, such as corn, wheat, dates, hay, grapes and tobacco, have been sun-dried to remove their moisture, or water content. This is necessary either to improve their quality, or so that they can be stored for longer periods of time. Many crops will rot very quickly if not dried before storage. Some crops like hay, are left in the field for drying. Raisins, which are dried grapes, and dates are usually placed in flat pans to be sun-dried to preserve them and also to make them taste better. Other crops are dried by artifical heat. But recently, sun-drying of crops has increased due to the fuel shortage, and the interest of people in returning to more natural ways of doing things.

From primitive times to today, human beings have also used the sun to get salt from the ocean. A flat area surrounded by dikes or earthen walls is flooded with salt water. The sun's heat then *evaporates*, or extracts the water. The salt is left behind on the ground to be collected, cleaned and sold. In the United States today, the large-scale industrial production of solar salt is about two million tons annually.

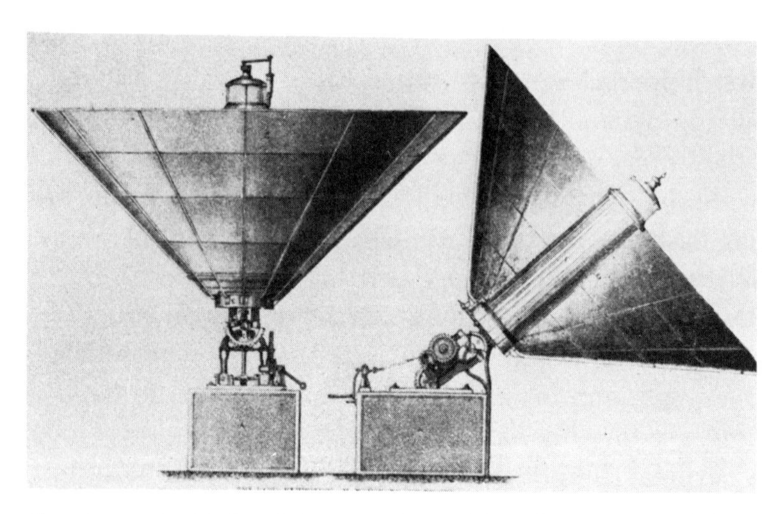

This is a Multiple Tube Sun-heat Absorber invented by a Frenchman, Mouchot, in 1878. It was used in a 176-day test of distilling water.

Another use of the sun's power with salt water is to convert it to fresh water by using solar stills. A *still* is a device in which a liquid, such as water, is heated until it starts to *vaporize*, or turn to steam or mist, which is collected and turned back to liquid again, leaving behind the impurities and salts.

As early as 1872, a large solar still was built at a mine in Chile to provide fresh water from salt water. It worked for over 40 years. Slanting roofs of glass over troughs full of salt water transmitted the sun's rays to heat the water. As the water evaporated because of the heat, the vapor rose to the glass roof, condensed into water and ran into collectors. This was fresh water—the salt was left behind in the troughs. This method of providing fresh water from salt water is used in several places in the world today where fresh water is scarce. On a good sunny day, a solar still can produce one pound of water for every square foot of the trough area.

Early settlers in the United States often used the sun's power to heat water. The water was put in a black-painted or dark iron tank, to better absorb the sun's heat. Then the tank was set in the sunlight for the day. By evening there would be hot water for washing.

An ancient and still widely used solar dryer is the clothesline, on which wet clothes are hung to dry in the sun's heat.

In France, in the 1600s, the first of a number of solar mechanical power systems was developed. These machines used sun-heated air or water to drive engines to power pumps and other types of machines.

A sun-power plant built by another Frenchman, Pifre, to heat a boiler to make steam for running a printing press.

In the United States, two men in particular tried to harness the sun's energy for power. They were John Ericsson, the designer of the famous Civil War ship, the *Monitor*, and A.G. Eneas.

John Ericsson claimed to have built the first steam engine to be driven directly by solar energy between 1868 and 1870. Between 1870 and 1883, he built several other solar engines.

Ericsson's sun-power plant of 1883, built to drive a steam engine.

An Eneas sun-heat absorber built in Arizona in 1903 to pump irrigation water.

A. G. Eneas designed and built three solar-powered steam engines to pump water to irrigate lands in the west from 1901 to 1904. From his design, a group of Boston investors built a fourth machine which was installed at the Ostrich Farm in South Pasadena, California. This machine ran successfully for a number of years.

During the 1900s, solar-driven machines continued to be designed and built all over the world. But cheap coal, and later oil and gas, were already on hand, and so no one tried very hard to develop solar energy.

USING SOLAR ENERGY TODAY

Today, due to the shortage of fossil fuels and their rising costs, there is an ever-increasing interest in the use of solar energy. Because of this interest, individuals and industries are experimenting with, building, selling and installing solar devices such as solar heaters. The most popular use of solar energy today is for heating water and buildings.

Using a device such as a solar heater is using the sun's energy *actively*. Using the sun *passively* is more common. Montezuma's Castle used the sun's power passively. There are many ways to use solar energy passively—the way a house is positioned, the kinds and amount of windows it has, the trees around the house to block the sun's heat in summer when they are in leaf, and so on.

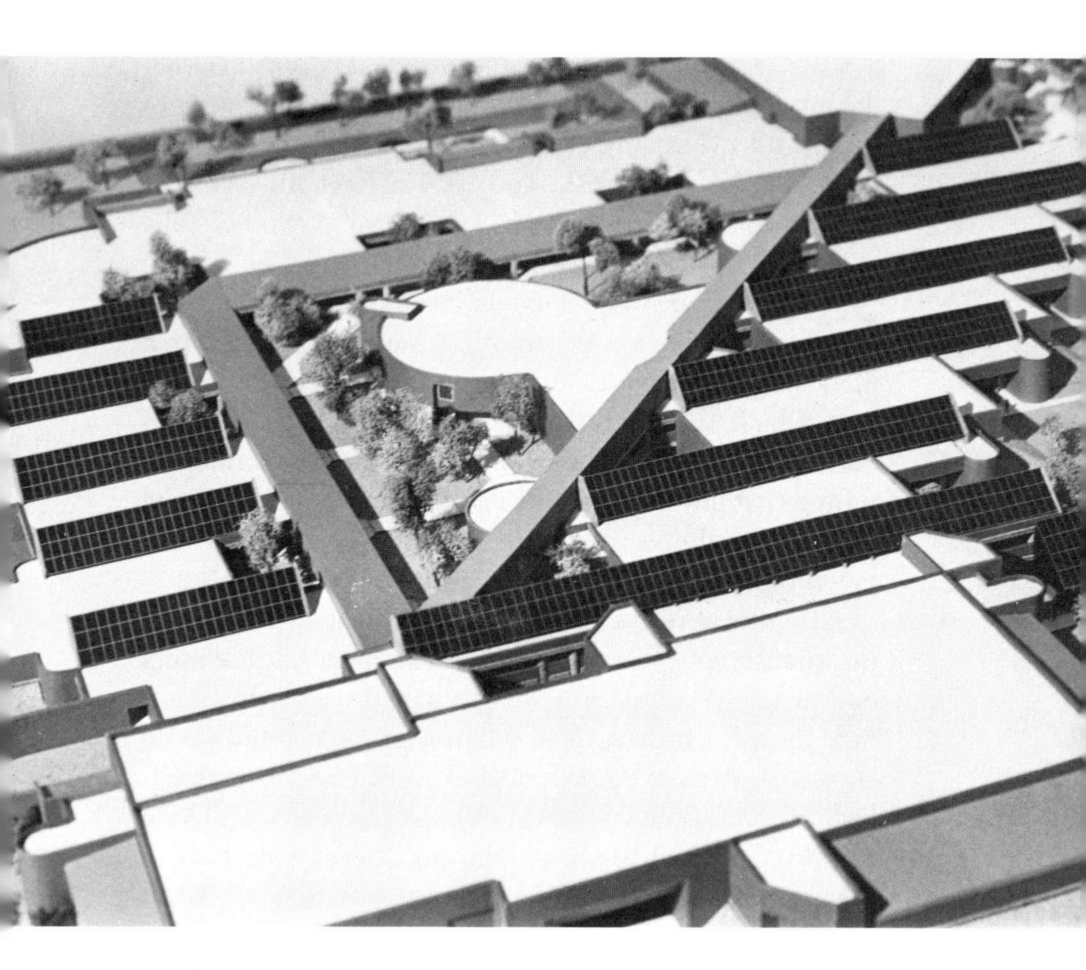

A solar heated and cooled high school in Tempe, Arizona.

Usually, solar equipment processes the sun's energy in one of three ways. One way is to *collect* it. The device simply gathers the sun's rays without doing anything else to them. Another way is to *concentrate* it. Concentrating the sun's rays through a magnifying glass onto a piece of paper will eventually cause the paper to flame up. The third way is to *convert* the sun's power to another form of energy. An example is the use of solar or photovoltaic cells to convert the sun's rays to electricity.

Most solar hot-water heaters use a flat-plate collector, an active device. The collector is usually a flat box with a piece of glass or plastic over the top. The bottom of the box, which is generally metal, is painted flat black to better absorb the sun's rays. The water flows through tubing that flows into an insulated water tank for storage until used. Often a solar water heating system can be put together with a few tools and readily available supplies.

The collectors used in buildings are usually located on the roof to get the most sunlight. They gather the sun's energy to heat water or air to circulate through the building. In order to have heat when there is no sunlight, a storage unit must be provided. This can be a large insulated tank, like a huge thermos, for storing hot water, or a bin of rocks which are heated by sun-warmed air. Both of these systems need electrical fans or pumps to circulate the heated air or water through the building. Often a conventional heating system is also needed to back up the solar system in case of long periods of cloudy days.

A commercial flat plate collector with a glass top and tank attached (rear).

A concentrating collector, another active solar device, is usually used to heat air or water for large buildings. These solar devices use a reflector to concentrate the sun's rays on the tubing filled with either air or water. They usually have a built-in tracking device to follow the sun's path in the sky and increase the amount of solar radiation that can be collected all day long.

Buildings may be air conditioned, too, by using solar energy and some of the same equipment that is used to

A commercial concentrating collector.

A solar heated home in Denver, Colorado.

generate solar heat. There are several ways to air condition a building using solar energy. One of the most popular is to use the sun's heat to power an engine that runs the air conditioner.

Solar hot water heats a liquid with a low boiling point, like ammonia or Freon. Vapor is generated and this is used to drive an engine. The engine operates the device to create cold air.

A solar oven with mirrors as the reflecting surface.

One of the most widely used solar devices today is the solar cooker. They may be bought from commercial companies, or made at home. Basically, the solar cooker is a box or cone-shaped device that is lined with mirrors or polished metal or foil materials that can reflect and magnify the sun's rays. In commercial cookers, a glass dome is often in the center of the reflector to trap the heat of the concentrated sunlight.

The cooker gathers heat from the sun and concentrates it in a small area, resulting in high temperatures. Over 500° F can be reached, which can cook any food.

The temperature of the air has nothing to do with the temperature that the cooker can reach. It is the sun's rays that generate the heat.

Some cookers can hold several pounds of food to be cooked and on a good day can cook a twelve pound turkey in four hours.

Probably the most exciting solar device to watch is the solar furnace. A solar furnace is a very complicated active device for concentrating solar radiation on very small targets to produce very high temperatures. Most solar furnaces have been built to carry out experiments. Some solar furnaces have used five-foot diameter searchlight reflectors as their base. Others, such as the one in the French Pyrenees Mountains, uses over 20,000 square feet of curved glass reflectors as the base to gather the sun's heat.

A large solar furnace built in Algeria. Besides carrying out experiments, it was used to produce fertilizer by the fixation of nitrogen in the air.

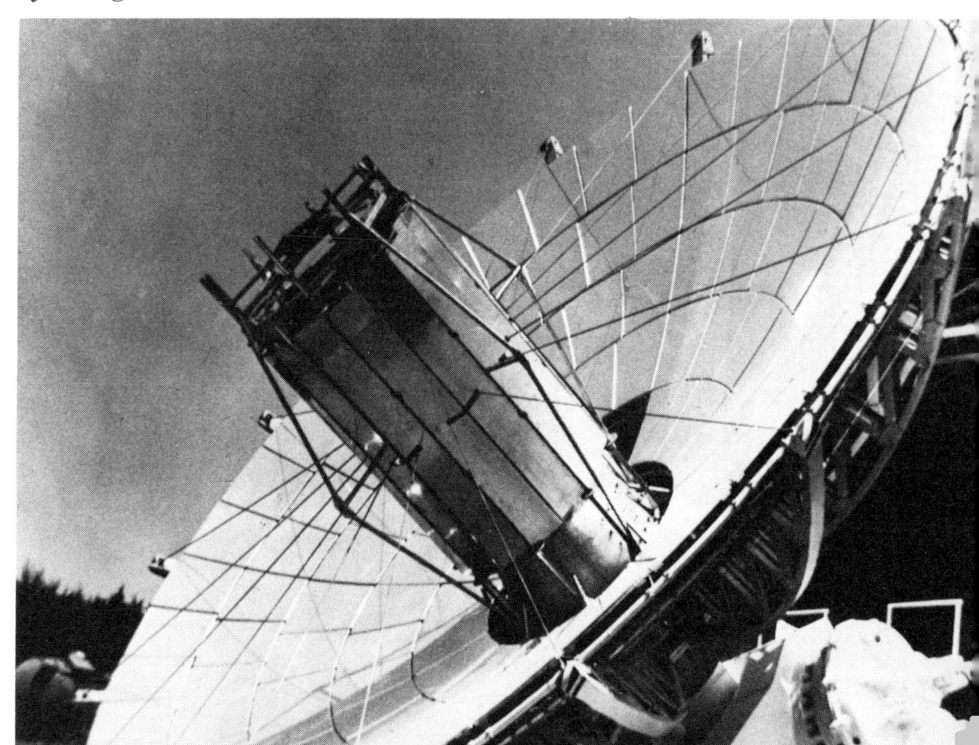

The solar furnace in the French Pyrenees generates temperatures of 6,300° F. One minute of this heat can melt a foot-wide hole in a piece of ⅜-inch steel plate! Because it generates such high temperatures, the solar furnace must have some sort of shutter device to control the reflected solar rays.

Another current use of solar energy is in solar or photovoltaic cells. These are solid-state devices—they have no moving parts—that convert sunlight directly to electricity. Solar cells at their best convert from 12 to 14 percent of the solar energy they receive to electrical energy. If protected from damage, they last a long time.

The solar cells are usually made of silicon crystals, which cost a lot of money to produce. There is a possibility of making solar cells cheaper or more efficient through the use of special dyes, concentrators, or by using materials other than silicon.

When they are cheaper, solar cells will be more widely used. However, even though still high-priced, there is increasing use of solar cells today. Solar cells are used in the

A microwave communications station powered by solar cells in an isolated area of Arizona.

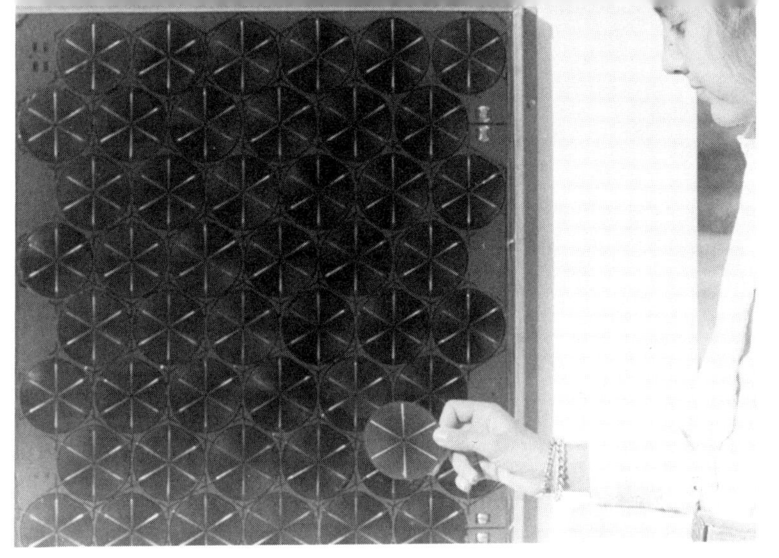

A panel of solar cells used in an exhibit to power a tape player. Such a panel costs about $1800.

space program to provide electrical power for satellites, where the need for light weight is more important than the high costs—hundreds of thousands of dollars. Solar cells work well in space, because, if positioned correctly, they can receive the sun's rays constantly. Solar cells are used in other special cases where it is impractical to get electricity from normal power sources.

Radios, telephones, lighthouses, navigational buoys, and even wristwatches and calculators are being powered with solar cells. Recently, a solar-powered radio transmitter was used by scientists in Australia to track the movements of a saltwater crocodile.

Closeup of a solar cell.

EXPERIMENTING WITH SOLAR ENERGY

Caution: *Experimenting with the sun can be dangerous. Burns to the skin can be caused by concentrated sunlight. Prolonged staring at the sun can cause blindness. Do not look into the sun or at the focal spots of sunlight. These spots produce extremely bright light and can be harmful to the eyes if you look at them for long periods of time. Remember, you don't feel any pain! Never look at the sun directly through any telescope—you could be blinded in a fraction of a second.*

MEASURING THE SUN'S ENERGY

As a class project, use a thermometer to measure the temperature of the interiors of several automobiles on a hot sunny day in a school parking lot. You can use any kind of thermometer, but be sure to get the car owner's permission first! Measure the temperature of the interior of several cars, some with the windows rolled up and some with the windows rolled down. Measure the temperature on the front seat of each car for ten minutes. What did you find? Were the cars with the windows rolled up hotter than the ones with the windows rolled down?

The closed automobile does not allow the solar energy to escape. The air inside the car is heated by the sun and also by the reradiation from the hot parts inside the car. *Reradiation* is the sun's energy which is trapped and does

Melissa, Darren, and Alfred are checking the temperature of the inside of the car.

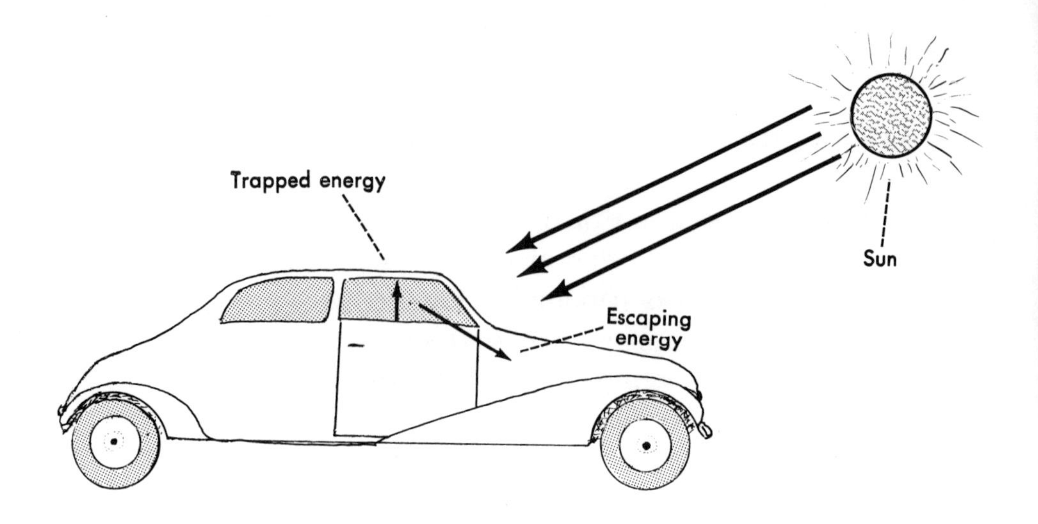

not escape from inside the car. This trapped energy continues to reradiate the air and provide additional heat. Therefore, the inside of the closed car is heated both by the direct rays of the sun and the reradiation from the inside of the car

The attic of a home is like the interior of a car. The sunlight absorbed by the roof makes the attic very hot in the summer. Many attics and automobile interiors can reach temperatures as high as 140° F in direct sunlight during the summer time.

You can use the sun's heat to save energy in your home. Raise the window shades or venetian blinds in the winter time, so the sun's rays can shine in. The room with a sunlit window should be warmer than the other rooms. In the summer, lower the shades or blinds to keep out many of the sun's rays. The room will be cooler.

BLACK HOTTER THAN WHITE?

Find a black cup and a white cup or two paper cups, one of which can be painted black and the other white. Place a thermometer in each cup and set them in the sunlight. Every 30 minutes take the temperature of each cup. Do this four times, and write the temperature down each time. Did you find that the black cup was hotter than the white cup?

Try the same experiment with water added to the cups. Cover the cups with clear plastic wrap. Is there a difference in temperature with water in the cups? Is there a difference between the white and black cups?

Now Melissa, Darren, and Alfred are using their thermometers to check temperatures of the white and black cups.

Take two shoeboxes and paint the inside of one black, or line it with black paper. Line or paint the other box white. Now cover the top of each shoebox with clear plastic wrap. Cut a slit in the side of each shoebox and insert a thermometer so the temperature can be read through the plastic. Set each box in a window which receives direct sunlight. Read the thermometers and record the temperatures four times, waiting twenty minutes between recordings. Did you find the black box hotter than the white one?

The more light energy an object absorbs, the warmer the object gets. This is because light energy is transferred into heat energy when it is absorbed or trapped. The color of the object influences the amount of light absorbed. White or light colors reflect the light rays and so do not trap many of them. Black or dark colors trap many light rays, and so much light is transformed into heat energy.

On a hot summer day you would be more comfortable wearing white-colored clothing than black. Cars with black interiors are also hotter than those with white interiors. So, if you live in a hot climate and want to be cooler, wear light-colored clothes and drive in a light-colored automobile.

MAKING SUN TEA

The common way to make tea is to boil the water and pour it on a tea bag or loose tea leaves. But here's a way to use the sun's energy to make tea. Take a gallon jar of water and add nine tea bags. Set the jar in direct sunlight. In several hours, heat from the sun and reradiation make sun tea. Leave in sunlight until the desired strength is reached. Sun tea is popular in desert regions.

Matt and his sun tea.

CONVERGING LENSES

Using a magnifying glass you can concentrate or *converge* the sun's direct rays into one spot. Lenses which do this are often called converging lenses.

Using a converging lens, focus the sun's rays to one spot on a board. This spot is called the *focal point*. Since all the light rays are converging to the focal point, the heat is also concentrated to this spot. BE CAREFUL NOT TO LOOK AT THIS FOCAL SPOT FOR ANY LENGTH OF TIME OR HAVE THIS CONCENTRATION OF LIGHT TOUCH ANY PART OF YOU! Soon the board will start smoking and burning at the focal point of the light. With practice, you can burn your name into a board using the converging lens.

Try the lens on dry leaves and paper. Do they burn?

Focus the light spot on a small glass or test tube of water. Can you make the water boil? BE CAREFUL SO YOU DO NOT GET BURNED BY THE BOILING WATER. Hold the glass or tube with asbestos gloves or tongs. Remove from heat immediately after the water starts to boil.

Try aiming various glass bottles in sunlight to see if the glass acts as a converging lens. Some parts can. First, hold the bottom of the bottle in the sunlight. What is the result? Now try the top and sides. Are any parts of the bottle like a converging lens.

Melissa is showing Alfred and Darren how to woodburn using a magnifying glass.

Mike is checking to see if part of this bottle can be used as a magnifying glass.

THE GREENHOUSE EFFECT

Have you ever seen a greenhouse? Have you seen the plants growing inside it? The sunlight comes in through the glass and it is trapped. Remember what sunlight did to the interior of the car. As the rays are trapped, they raise the temperature of the air inside the greenhouse. Any surface covered with glass or plastic and facing the sun, traps heat energy. This is called the *greenhouse effect.*

Solar collectors use the greenhouse effect to store solar energy. Solar collectors are usually blackened plates covered with a clear layer of insulation, which allows light through but prevents air currents from cooling the plate.

To make a collector, paint the interior of a shoebox black. Place a jar of pebbles, a jar of sand, a jar of water, and a jar of dirt inside the box. Baby food jars work nicely. Place a thermometer in each jar. Cover the shoebox with a clear plastic cover, and set in the sunlight. Record the temperature of each jar every half hour for four hours. Now move the box out of the sun and in a half hour take a last reading. Which jar was the best collector of energy? Which jar had the highest temperature after it was removed from the sunlight?

These types of experiments are what scientists carry out to build solar collectors. From your experiment, what do you think would be a good solar collector?

Mike sets up the solar collectors.

Steve is covering the solar collectors with food wrap.

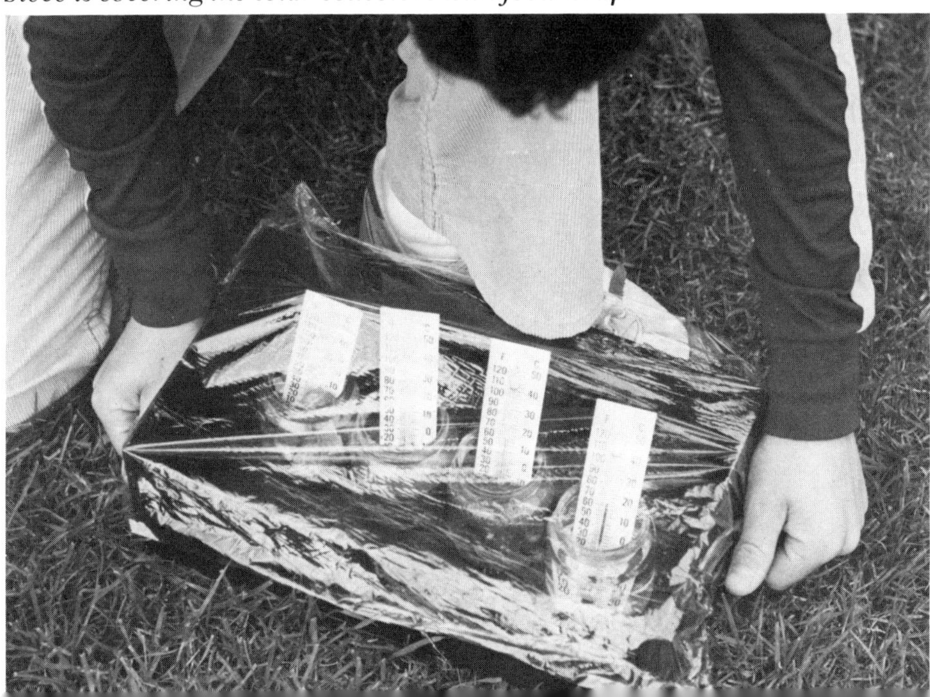

A SOLAR COOKER

Would you like to cook a hot dog and not use any electricity or fire? It can be done by using the sun as the source of energy. You can build a solar hot dog cooker with an aluminum pop can, aluminum foil and a piece of clothes hanger wire. Find an aluminum pop can and carefully cut out a five-centimeter-wide (about two inches) section with scissors. *Do not use sewing scissors or it will ruin them.* Fold the ends inside and bend down all sharp sides. Line the inside of the can with aluminum foil. Now take a piece of wire 20 centimeters (about eight inches) long—clothes hanger wire is good. Guide the wire through the can and the hot dog. Set in direct sunlight and watch your

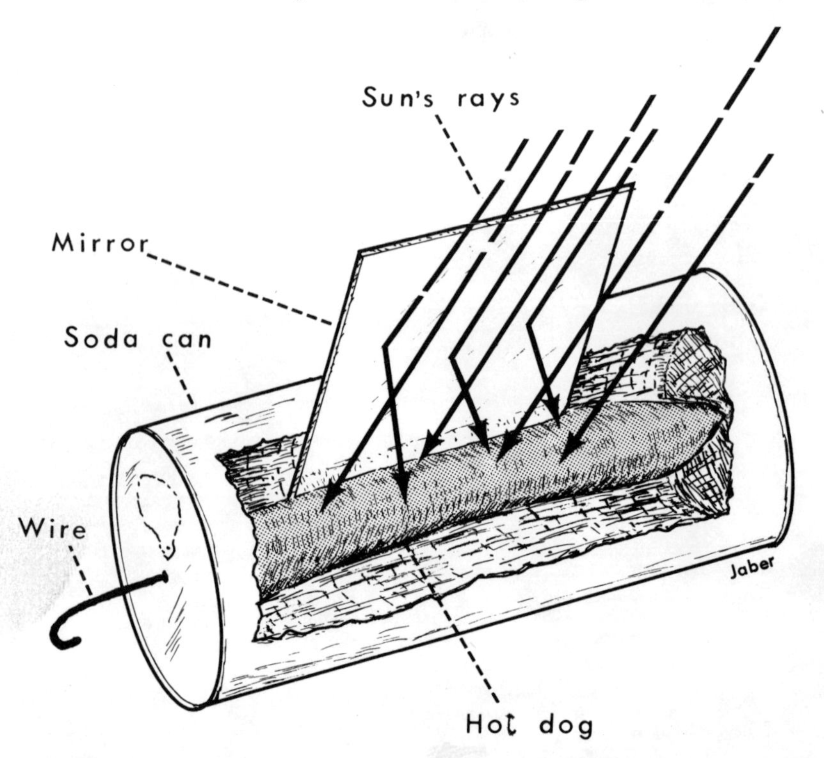

Solar hot dog cooker.

Steve building a hot dog cooker.

hot dog cook. You can speed up the process by placing a small rectangular hand mirror inside the can. Focus the mirror to reflect the sun's rays on the hot dog. The solar cooker collects the reflected sunlight and concentrates it to produce enough heat to cook the hot dog.

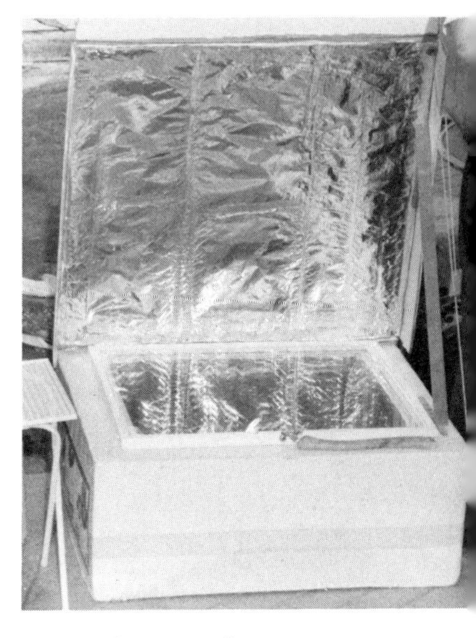

On the left, a solar cooker, 1876 model. And, today, a solar oven made from two cardboard boxes, with aluminum foil as the reflecting surfaces and a double pane of glass to trap the heat.

CROOKES RADIOMETER

Crookes Radiometer detects the presence of heat radiation. It is a tube from which most of the air has been removed. There are four vertical vanes mounted on a vertical axis. The vanes are black on one side and either white or polished on the other. The vanes spin freely.

Find a Crookes Radiometer in a novelty, hobby or science materials shop. Set it in the sunlight and watch it turn. How do you think it works? Scientists explain that, as the sun's energy strikes the vanes, the black surface absorbs more energy than the white or polished surface. This raises the temperature of the black surface above the temperature of the white or shiny surface, causing the air molecules of the black surface to move more quickly. The rebounding air molecules against the sides of the vanes cause them to turn.

Crooke's Radiometer.

SOLAR FURNACE

Be careful when doing this experiment. Do not look at the projected image. Keep a pail of water on hand to put out any fire. This experiment must be done outdoors.

You will need a Fresnel lens for this. Fresnel lenses can be found in most hobby shops. They are converging lenses. In other words, the Fresnel lens focuses the sun's rays from large regions to one concentrated spot. The Fresnel lens can produce temperatures of up to 2000° F and the hot spots are as bright as the arc of a welder.

Since it is difficult to hold the Fresnel lens steady and to one spot for any length of time, build a stand to hold the lens. It should rotate so it can be focused to the direct rays of the sun. See the diagram on the next page for construction details.

The solar furnace-cooker set up to cook.

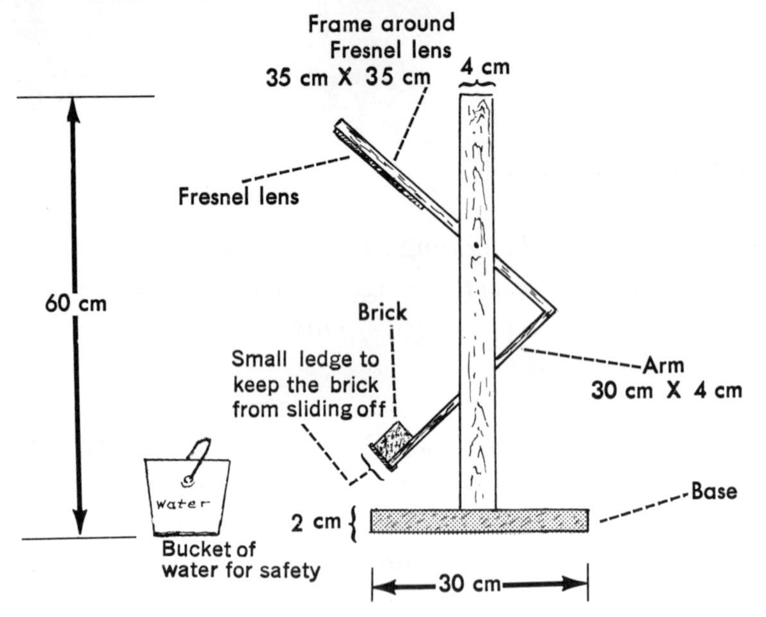

Diagram of parts for you to build your own furnace-cooker.

Using a piece of newspaper, find the focal spot by moving the paper until a bright spot of light is seen on it. Once the newspaper is placed at the focal point of the Fresnel lens, it should start to burn within seconds. Now, find a brick and hollow out a 3-centimeter hole (a little more than an inch) in it. Place the brick on a ledge at the end of the arm of the Fresnel lens stand at the focal point. Then

The solar furnace-cooker ignites newspaper in just a few seconds. BE CAREFUL.

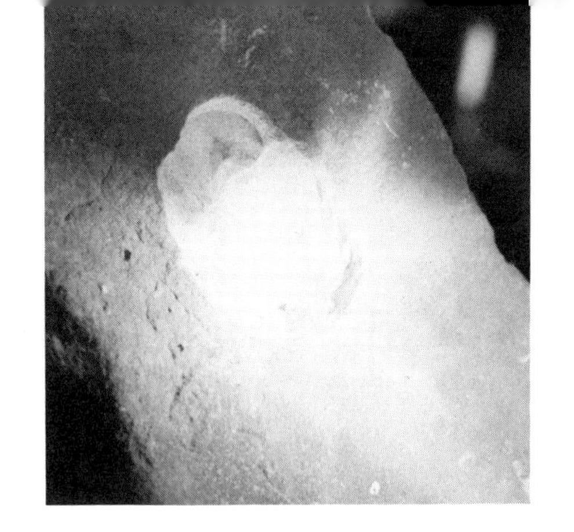

Melting solder in a brick crucible (a special vessel that withstands high heat), using the fresnel lens.

adjust the Fresnel lens to focus the concentrated solar energy to the hole of the brick. The brick and heat from the Fresnel lens act as a furnace. Place objects such as solder, lead and junk jewelry on the brick furnace to see the results of the intense heat.

You can make a solar cooker by replacing the brick on the arm with a can of water. Place an opened can of soup (or whatever can you are cooking) in the larger can of water. As the water boils, the soup will cook.

Melissa, Alfred, and Darren are ready to eat their sun-heated soup.

Building the solar still.

THE SOLAR STILL

In desert survival courses, the Boy Scouts and other groups have used the concept of the solar still as a means of collecting water. Here is what they do:

Dig a hole about 50 centimeters (nearly 20 inches) deep and about 100 centimeters (40 inches) across. Center a cup or glass at the bottom of the hole. Now cover the hole with a clear plastic sheet. Place dirt around the rim of the hole to keep the clear plastic in place. Put a small rock on the plastic cover to sink it to a low point in the middle over the container. The cover should reach to about 4 centimeters (a bit more than 1½ inches) from the top of the container. As the solar heat is generated under the clear plastic sheet, it causes the moisture in the ground to evaporate. As the moisture evaporates, it rises, hits the plastic and changes to water. The water accumulates, slowly runs to the bottom of the plastic, and drops into the container.

If you can't dig a hole such as has been described, a solar still can be made with two jars. You will need a jar that fits easily inside a larger jar, and it must be about 6 centimeters (2½ inches) shorter in height. You will also need a small piece of clear, wetable plastic.

Set the smaller jar inside the larger one. Pack dirt between the jars to near the top of the small jar. Wet the dirt with water. Fasten the clear plastic to the top of the larger jar with a rubber band. Place a small rock on the plastic, so it sags down into the smaller jar. Put the jars in the sun. Water will collect inside the smaller jar by the same process of evaporation, condensation, and accumulation.

Completed model of a solar still.

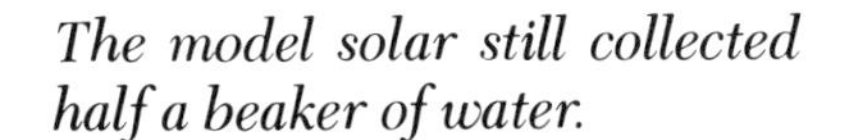

The model solar still collected half a beaker of water.

THE SOLAR ENGINE

There are several kinds of solar engines. Some of them are based on the fact that metals and most liquids *expand*—get larger—when hot, and *contract*—get smaller—when cooling off. But the most successful solar engines in the past have been those that used steam to work.

Solar steam engines work the same way as cookers to get steam. Sunlight is focused on a boiler. The water is raised to the boiling point, and the steam that results is channeled off in pipes. Under pressure the steam is able to turn wheels or do other work.

A solar steam engine was built and put into operation at a show in Paris, France in 1878.

A man in New Mexico invented a solar steam engine that pumped water 20 feet up into a large tank. He then allowed the water to run down as a small waterfall, hitting a paddlewheel as it fell. The paddlewheel turned another wheel that generated electricity.

Up to now, solar engines have not been able to compete with other kinds of engines. This has been because they are expensive to build and to maintain, and they cannot run all the time, since they depend on sunlight for their energy.

The liquid or metal expansion engines are much more complex, and yet they may be used more in the future because they are more efficient.

Steve, Matt, and Mike are setting up the solar engine.

SOLAR ENGINE EXPERIMENT

Here is a simple expansion solar engine. This is a more advanced experiment, which can be done in the classroom or with adult help.

NASA (National Aeronautics and Space Administration) discovered the concept of the expansion solar engine in the late 1960s. The extendable booms of satellites were made of thin metals. As the booms came in contact with the sun's heat, even though it is very cold in outer space, the booms would twist about wildly. Dr. Richard Beam of NASA discovered that these booms were a crude form of solar heat engine—an expansion engine. The NASA experiment* is a model of this engine.

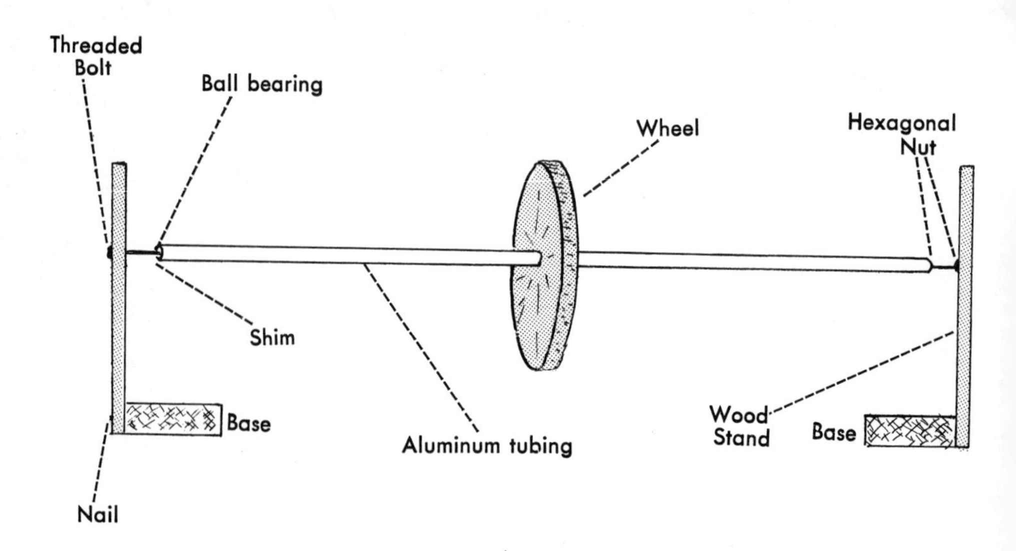

Diagram of solar engine.

**Power Directly From the Sun*, Public Affairs Office, NASA - Ames Research Center, Moffett Field, Calif. 94035

You will need the following:

Masking tape

Duco or rubber cement; epoxy glue

2 ball bearings, plain (without shields), ¼-inch bore, ⅝-inch outer diameter. The bearings must spin freely.

1 aluminum tube, ¾-inch outer diameter, 6 feet long. Paint tube flat black with a very thin coat. (A thick coat will prevent rapid heating and cooling.)

5-pound box of patching plaster.

2 machine screws, 2 inches long, fully threaded.

2 flat washers.

2 hexagonal nuts.

4 pieces of 1-inch by 4-inch lumber, 9 inches long. Make a ¼-inch hole, about 1 inch from the end, in the center of the lumber. Nail into stands as shown in the sketch.

2 pieces of 0.012 brass shim stock ¼″ wide, 1 15/16″ long. The shim stock will take up the space between the bearings and the inside diameter of the aluminum tube so that the bearing will fit tightly into the tube.

1 piece of manila folder material 3″ by 4 ⅞″

1 plastic lid from a 3 lb. coffee can

Modeling Clay

Candle wax

Directions for Assembling Solar Engine

1) First, form the manila folder around some ½″ rod.

2) Tightly roll the folder around the aluminum tube, gluing the paper to itself in several spots (Duco or rubber cement works well). Hold down the final edge with masking tape.

3) Rub candle wax on the paper tube, then remove from the aluminum tube. The paper tube will be used as an inner form.

4) Roll some modeling clay into a tubular shape a bit smaller than the paper tube. Push it through the aluminum tube. Stand the clay-filled tube on a table and press down on the clay with the thumb so that the clay will expand in diameter to solidly fill the paper tube. (The clay will keep the paper tube round and also be used to fasten it to a plastic coffee can lid which will be used as a flywheel.)

5) Cut 2 strips of manila folder material 9 13/16″ by 3″ and form them into a hoop with masking tape. The hoop should just go around the plastic lid.

6) Fasten the paper tube to the center of the plastic lid by pressing hard on the modeling clay to make a strong joining. Hold the lid up to the light to make sure the paper tube is well centered (plastic lid has a thicker spot in the center which is helpful in locating its center).

7) Place the finished mold on a table.

8) Pour 3 cups of water into a container and add patching plaster (stirring frequently) until the mixture has the consistency of soft pudding. Only about ⅔ of the patching plaster will have been used. Pour this into the mold previously made. Level the mixture to a smooth surface. Let harden about 1 day before stripping the mold from the flywheel. The flywheel can now be glued to the center of the aluminum tube with epoxy glue. The epoxy will take about 1 day to harden.

9) Put the shim stock ring partway into the aluminum tube and push in the bearing and ring so that they are even with the end of the tube. Don't worry if the bearing and ring are pushed in too far: just pull them out and try again. Withdraw the stands until a few screw threads show (the bearing should not touch either nut). With the tube assembled on the stands, the engine is complete except for balancing.

10) One spot on the flywheel will be heavier than the rest, and the flywheel will quickly turn until this spot is on the bottom. You can mark this spot with a pencil and verify that no matter to what position the flywheel is turned, the pencil spot will stop at the bottom.

The idea is to remove this heavy spot by carving away the bottom material (use a pocket knife). Carve and test. As perfect balance is approached, the heavy spot will move more slowly toward the bottom, until finally the flywheel shows no preferred spot for the bottom position. The balance is then perfect. If too much material is carved from the bottom, this region will now be lighter than the rest and always stop at the top. Don't worry: just treat the new lowest spot as being too heavy and proceed to carve away from it. (Only remove less material this time!)

The amount of heat required to operate the engine, and how fast it will run, depend on how free the bearings are and how well balanced the flywheel is. The engine can be operated indoors with the aid of a heat lamp or two, or outside from the sun.

Heat should be applied to the top region of the aluminum tube near the flywheel—(the portion near the bearings is ineffective). Heat applied to the bottom works in the wrong direction—it will slow the engine down. This is a good point to remember if the engine is placed on a hot surface to be operated by the sun. Cool asphalt by wetting it down, for example.

The materials for this solar engine were chosen for their availability and low cost, not for performance. A thinner tube of a higher strength material would greatly improve the power and efficiency of the engine.

SOLAR CELL EXPERIMENT

Solar cells convert light energy directly into electrical energy. Solar cells, often called photovoltaic cells, are made up of single crystals of silicon. Silicon is a nonmetallic, sandlike element. Scientists explain that a solar cell works this way: As the silicon crystals are exposed to solar radiation, the radiation is absorbed. As the radiation is absorbed, it causes the silicon atoms to give off energy—electricity!

Purchase a solar cell from a hobby shop or borrow one from the science lab. Using a night light bulb or a Christmas tree light, the solar cell, and insulated electrical wire, experiment on getting the light to come on. Connect the cell with the bulb with the wire, one end on the metal bottom of the bulb, and the other end on the side of the metal. With the use of solar cells, electricity can be generated directly from sunlight. Visit a hobby shop to see solar-powered model airplanes, cars and solar radios.

Checking a small solar cell with a meter to see how much electrical current it can generate.

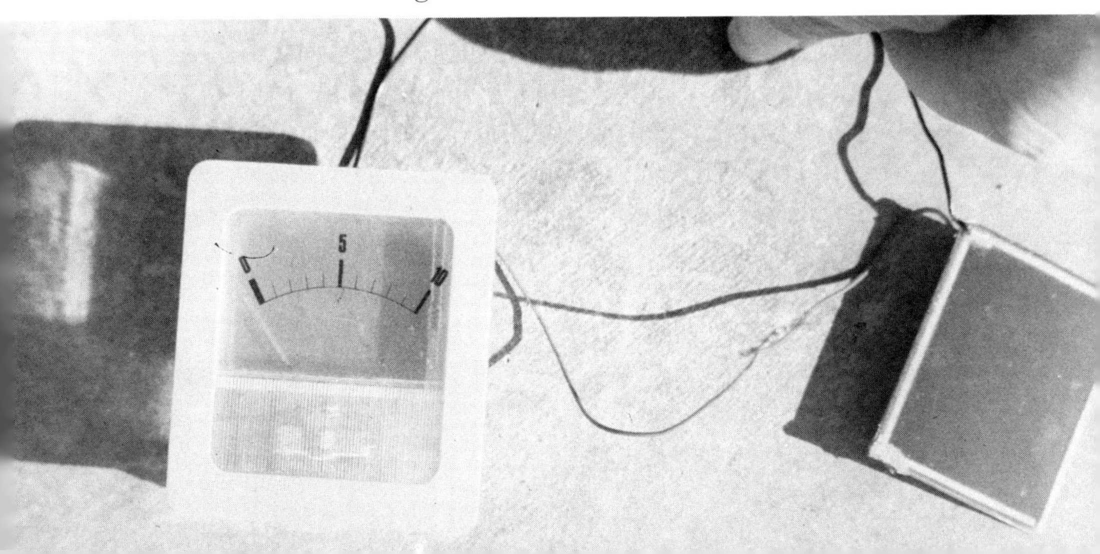

THE FUTURE AND SOLAR ENERGY

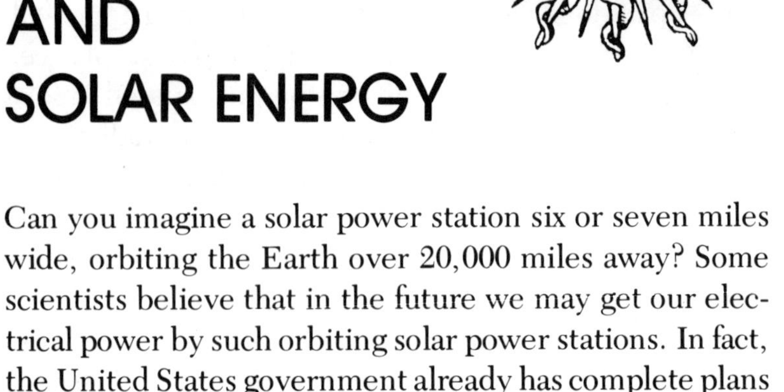

Can you imagine a solar power station six or seven miles wide, orbiting the Earth over 20,000 miles away? Some scientists believe that in the future we may get our electrical power by such orbiting solar power stations. In fact, the United States government already has complete plans for construction of such a power station. The station would cost billions of dollars to build.

These stations would work by using panels made of solar cells, miles in length and width. Except for short periods of time during solar eclipses, the panels of solar cells would receive constant sunlight. The power plant would move at the same speed as the earth turns and would therefore be positioned in one spot all of the time. This is necessary so that the electricity generated by the power station could be sent down to one place on the earth's surface for distribution and use.

But then the problem arises of how to get the electricity to earth. One of the great miracles of science is that all kinds of energy forms, such as light, heat, sound, and electricity can be converted from one form to another. Electricity can be converted to high energy radio waves called *microwaves*, beamed down to earth, and then converted back to electricity by the earth power station. Then it can be sent by regular powerlines to wherever it is needed.

Someday, solar cells may be used in space to generate large amounts of electrical power.

Solar cells may also be used in large quantities on roofs or in windows to generate electricity for buildings. We have the capability to do this today, but we need a great reduction in price of solar cells before it is practical.

Another way that the sun's power may be trapped in the future is by using sun-warmed seawater to generate electricity. Scientists have already developed models to do this.

The machine that would convert the heat of seawater into electricity is called a thermal-powered electrical generating plant. The electrical generator would probably look like a tube and it would be over 1,500 feet long. The top of this tube would be so large that it would house the crew needed to work it. One hundred feet below them would be the inlet pipe and generators. The plant would be located in an area where the surface temperatures of the sea would be around 80°F, perhaps off the coast of Florida.

The warm surface seawater would be pumped through the inlet pipe to heat a liquid chemical such as ammonia, which has a very low boiling point, lower than the boiling point of water. The warm seawater would cause the ammonia to boil, turning it into vapor. The vapor would drive a turbine to generate electricity. To cool the ammonia vapor to liquid form again so it can be reused, cold seawater from deep down in the ocean would be pumped from the bottom of the tube. Since ocean water stays at a fairly constant temperature, the process could be continuous.

On a much smaller scale, a solar pond can be used. A *solar pond* is a hole in the ground filled with salt water.

It has been known for a long time that saltwater ponds have the ability to receive and store solar energy. The solar pond has a bottom layer of around 20 percent salt water and then several layers of less salty water, with a layer of fresh water on top. The bottom layer of salt water is the heaviest, due to the higher percentage of salt. Each layer of less salty water will float on top of the more salty layer below it, if added carefully. The less salty layers and the freshwater layer act as insulation. The sun's heat can penetrate to the bottom layer, and the insulating layers of water prevent the stored heat at the bottom from escaping. Then, by circulating water through a pipe at the warmest level of water, heat is extracted and can be piped to a building.

A pond with the same square-foot surface area as the building using it can probably provide a full winter's heat.

The solar pond may be the answer to farmer's needs to heat their houses, barns and drying bins for crops.

The world's first commercial, solar-powered, land-based, electrical generating plant is being built in the California desert. This plant, sponsored by the United States Department of Energy, is going to cost over 100 million dollars. The power station will cover an area of over 130 acres. It will have a field of over 1,500 computer-controlled mirrors to follow the sun. The mirrors will reflect and focus the sun's rays onto a boiler on top of a tower 283 feet high. The boiler's water, heated into steam, will drive an electrical generator on the ground and produce enough electricity for a town of 6,000 during the daylight hours. At night, when the solar generator is not running, the electrical system's power lines are automatically hooked up with power lines of other electrical sources. However, scientists are working on ways to build large storage batteries or other devices that will store solar-generated electricity in large amounts to avoid the need for other sources.

A solar process almost as old as the earth may play a big part in meeting our future energy needs, that is the generation of large amounts of hydrogen, a chemical element that makes a clean, efficient fuel. That solar process is photosynthesis. *Photosynthesis* is the process by which plants, using sunlight and their own chlorophyll (a green chemical substance), convert carbon dioxide and water to carbohydrates and oxygen. The carbohydrates then are stored either as fuel or food.

The reason animals cannot do this same thing is that only plants have the chemical substance *chlorophyll*, which is essential to the process of photosynthesis.

There have been some scientific breakthroughs in recent years in discovering exactly how photosynthesis works. In fact, experiments are now being carried out to create artificially the photosynthesis process. Scientists believe that in a few years, devices can be developed to use the sunlight, as plants do, to break down water into hydrogen and oxygen. The hydrogen could then be used as fuel. One great advantage to this process over other solar energy devices is that the process could convert sunlight to energy much more efficiently than most other planned solar devices.

At one time, humans were very dependent on plants for fuel. But plant fuel was abandoned in the 1800s because coal burned more efficiently. Today, experiments are being carried out using plants such as algae, seaweeds, fast-growing trees and grasses to see if better plant fuels can be obtained.

There is even one experiment to grow "gasoline" bushes. These are commonly called gopher plants. Their oil-bearing fruit could be harvested and used in place of petroleum which their oil is very similar to.

Solar energy is being used in space today to generate electricity for satellites but an even more fantastic use is being proposed by scientists—a spaceship fitted with huge solar sails, to be driven by the sun. The energy of the sun, which it radiates out at great speeds is called "solar

A drawing of the future—the design for a huge solar energy satellite, rectangular in shape and occupying almost 50 square miles.

wind." The solar wind can exert pressure on anything that gets in its way. This force of the sun's energy could be used to propel sailing spaceships across vast distances without their needing any other source of fuel.

Other scientists have proposed a new type of spaceship. It too, would have giant sails, but these would be paper-thin solar cells. The solar cells would generate electricity from the sunlight to provide power to run a series of tiny jet engines. These jet engines would use mercury as their fuel, letting mercury vapor escape from a nozzle to the rear, giving the ship forward thrust, the same as a jet plane.

The promise of solar energy to supply our power needs is great. It is said by some that within twenty years most buildings in the United States could be fitted with solar devices to provide for their heating and cooling needs. Some say that by the year 2025, over 80 percent of our energy needs can be met using the power of the sun.

There are still many problems to be solved. There are problems connected with generating solar-powered electricity, storing vast amounts of electricity for use at night when there is no sunlight to generate power, and making solar energy less costly.

Yet, the promise of the sun is great for a clean, non-polluting source of limitless power. By the year 2000, scientists believe that many of the technical problems in harnessing the sun's power will be overcome, and that we will have returned to the earth's basic source of energy, the sun.

INDEX

adobe, 12
air conditioning, solar, 25
algae, 60
aluminum, for building cooker, 40
ammonia, 25, 58
animals, 60;
  energy from, 10
Apache Indians, 15
Archimedes, 11
Arizona, 12
astronomers, 8
atmosphere, 10
automobiles, *See* cars
Aztecs, 11, 12

battery, solar, 59
Beam, Dr. Richard, 50
booms (of satellites), 50
buildings, 12, 20, 22, 24, 25, 59, 62. *See also* houses

California, 19, 59
carbohydrates, in photosynthesis, 60
carbon dioxide, in photosynthesis, 60
cars, 7, 31, 32. *See also* experiments: cars
chili, 16
chlorophyll, 60
coal, as source of energy, 10, 19, 60
color, 33, 34. *See also* experiments: color
converging lenses, 36. *See also* Fresnel lens
cookers, solar, 26, 27, 40, 41, 45, 48; directions for construction, 40, 41. *See also* experiments: solar cookers
corn, 15
Crookes Radiometer, 42
crops, 15, 59. *See also* individual names

dates, 15

earth, 8, 10, 56, 57;
  solar radiation to, 7
Egyptians, 11
electrical energy, 55
electrical power, 29, 56, 57
electricity, 22, 28, 29, 48, 55, 56, 57, 58, 59;
  solar-generated, 59, 61, 62
electromagnetic waves, 7
Eneas, A. G., 18, 19
energy, solar, 7, 10, 18, 19, 20, 22, 24, 25, 28, 32, 38, 42, 58, 61, 62;
  collection of, 22, 24
  concentration of, 22, 24, 26, 31
  conversion, 22, 36, 45
  current uses, 20-29, 60
  electrical, 28
  future uses, 56-62
  history of, 11-19
  sources of, 7
engines, solar, 18, 19, 25, 48-54;
  directions for construction, 51-54. *See also* steam engines
Ericsson, John, 18
experiments:
  cars, 31, 32, 34, 38
  color, 33, 38, 42
  concentration of solar radiation, 27, 30
  converging lens, 36
  homes, 32
  measurement of solar energy, 31
  photosynthesis process, 60
  plants (greenhouse), 38, 60
  reradiation, 31, 32
  solar cells (photovoltaic), 55
  solar cookers, 40, 41
  solar energy, 30-55
  solar engines, 48-54
  solar furnaces, construction of, 43-45
  solar stills, 46, 47
  sun tea, 35
Florida, 57
flywheel. *See* engines, solar; directions for construction
focal point, 36, 44, 45. *See also* experiments: converging lens; experiments: solar furnaces
food and solar energy, 9, 26, 27, 40, 41, 60
fossil fuels, 10, 20. *See also* coal; gas; oil
France, 17, 48
French Pyrenees Mountains, 27, 28
freon, 25
Fresnel lens, 43-45
fuel:
  production of, 59, 60, 62
  shortage of, 15. *See also* coal; fossil fuels; gas; oil
furnaces, solar, 27, 28, 43-45;
  construction of, 43-45. *See also* experiments: solar furnaces

gamma rays, 7
gas, as source of energy, 10, 19
generating plants. *See* power plants
generators:
  electric, 59
  solar, 59
gods. *See* sun: gods
gopher plants, 60
grapes, 15
Greeks, 11
greenhouse effect, 38. *See also* experiments: plants

hay, 15
heat:
  artificial, 15
  concentration of, 36, 38
  experiments, 31, 32. *See also* temperature, experiments
  from the sun, 7, 8, 11, 12, 17, 25, 27, 32, 58
  radiation, 42. *See also* Crookes Radiometer
heat energy, 34
heaters, solar, 20, 22
heliographs, 12, 15
Helios, 11

houses, 20, 59.
hydroelectric power, 10
hydrogen, 59

Indians, ancient, 12
infrared rays, 7
insulation, 22, 38, 58
irrigation, 19

jet engines, 62

lenses. *See* converging lenses
light, 7, 8
light energy, 34, 55
light rays, 7, 34

machines. *See* power systems
magnifying glass, 22, 36
mercury, 62
microwaves, 57
Miles, General Nelson A., 15
mirrors:
 computer-controlled, 59
 in heliographs, 15
 in legend of Archimedes, 11, 12
 in solar cookers, 41
Montezuma, 12
Montezuma's Castle, 12, 20
Morse Code, 15

NASA (National Aeronautics and Space Administration), 50
New Mexico, 12
nuclear energy, 10

ocean, 15, 58
oil, as source of energy, 10, 19
orange dwarf star, 8
oxygen, 60

petroleum, 60
photosynthesis, 59, 60
photovoltaic cells, 22, 28, 29, 56;
 uses of, 29, 55, 57, 62
pioneers, 12
plants:
 energy from, 10, 60
 growth of, 38. *See also* experiments: plants, individual names
power, from solar energy, 15, 17, 18, 62. *See also* electrical power
power plants, 56, 57, 59
power pumps, 17, 19
power stations, 56, 57, 59
power systems, 17

Ra, 11
radiation, 7, 24, 27, 29, 31, 55
radiowaves. *See* microwaves
reflector, 24, 27
reradiation, 31, 32, 35
rocks:
 in buildings, 12
 in heating, 22

salt, from solar energy, 15. *See also* salt water
salt water, 15, 16, 58
satellites, 29, 50, 61
seawater, obtaining energy from, 57, 58
seaweeds, 60
silicon crystal, 28, 55
solar cells. *See* photovoltaic cells
solar collectors, 38
solar dryer, 17
solar eclipse. *See* sun: eclipses of
solar pond, 58, 59
solar rays. *See* radiation
solar sails, 61
solar salt. *See* salt, from solar energy
solar system, 8
solar water heating system. *See* heaters, solar
space, 57, 61
space program, 29
spaceship, 61, 62
Spanish conquest, 12
stars, 8
steam engines, 48;
 history of, 18, 19, 48
 solar-powered, 19
stills, solar:
 construction of, 46-47
 definition of, 16
 uses of, 16, 17. *See also* experiments: solar stills
storage, of water, 22, 24
sun:
 as source of energy, 7, 10, 11, 12, 17, 18, 26
 characteristics of, 8
 eclipses of, 11, 56
 gods, 11. *See also* individual names
 worship of, 11.
sun power. *See* power, from solar energy
sun tea. *See* experiments: sun tea
sunlight:
 concentration of, 28, 30, 32, 33, 35, 41, 48, 56
 effect of, 38, 42, 55
 use of, 60, 62
Syracuse, Battle of, 11.

temperature:
 experiments, 30, 31, 32, 33, 38, 43
 from radiation, 27, 42
 of ocean, 58
 of solar cookers, 26
 of solar furnaces, 28
 of sun, 8, 12
thermometer. *See* temperature
Tezeatlipoca, 11
tobacco, 15
turbine, 58

ultraviolet rays, 7
United States Army, 12
United States Department of Energy, 59
United States of America, 15, 17, 18, 56, 62

vaporization, 16. *See also* stills, solar

water:
 circulation of, 58
 conversion of, 60
 evaporation of, 15, 16
 experiments, 35, 36
 fresh, 16, 17
 heating, 17, 20, 22, 25, 35, 36, 59
 pumping, 19, 48
weather, 8
wheat, 15
wind, solar, 61, 62

X rays, 7